KNOW YOUR DONKEYS & MULES

JACK BYARD

Know Your Donkeys & Mules © Jack Byard, 2010, 2012
All rights reserved.

First published in Great Britain as *Know Your Donkeys* in 2010 by Old Pond Publishing Ltd.
First published in North America in 2012 by Fox Chapel Publishing, 1970 Broad Street, East Petersburg, PA 17520, USA. Published under license.

No part of this publication may be reproduced, stored in a retrieval system, or transmitted, in any form or by any means, electronic, mechanical, photocopying, recording or otherwise, without prior permission of the copyright holder.

ISBN: 978-1-56523-614-1

Library of Congress Cataloging-in-Publication Data

Byard, Jack.
  Know your donkeys & mules / Jack Byard. -- 1st ed.
    p. cm.
  "First published in Great Britain in 2010 by Old Pond Publishing Ltd"--T.p. verso.
  ISBN 978-1-56523-614-1
  1. Donkeys. 2. Donkey breeds. I. Title. II. Title: Know your donkeys and mules.
  SF361.B98 2011
  636.1'82--dc23
                                                                2011017570

To learn more about the other great books from Fox Chapel Publishing, or to find a retailer near you, call toll-free 800-457-9112 or visit us at *www.FoxChapelPublishing.com*.

Note to Authors: We are always looking for talented authors to write new books. Please send a brief letter describing your idea to Acquisition Editor, 1970 Broad Street, East Petersburg, PA 17520.

Printed in China
First printing

Contents

- 6 Abyssinian
- 8 American Mammoth
- 10 American Spotted
- 12 American Wild Burro
- 14 Amiata
- 16 Andalusian
- 18 Asinara
- 20 Balear
- 22 Bourbonnais
- 24 Catalan
- 26 Cotentin
- 28 Cypriot
- 30 Encartaciones
- 32 Grand Noir du Berry
- 34 Irish
- 36 Kiang
- 38 Majorera
- 40 Martina Franca
- 42 Mary aka Maryiskaya
- 44 Mediterranean Miniature
- 46 Normandy
- 48 Onager
- 50 Pantelleria
- 52 Poitou
- 54 Provence
- 56 Pyrenees
- 58 Ragusano
- 60 Romagnolo
- 62 Sardinian
- 64 Somali Wild Ass
- 66 Zamorano Leones
- 68 Zedonk/Zonkey
- 70 What is a Mule?
- 72 Saddle Mule
- 74 Belgian Draft Mule
- 76 Mule Racing

Donkey Talk

ASS: Another term for a donkey.

BURRO: The American term for a small donkey.

JACK: A male donkey.

JENNY (OR JENNET): A female donkey.

COLT: A male donkey under one year old.

FILLY: A female donkey under one year old.

MULE: A male donkey crossed with a female horse.

HINNY: A male horse crossed with a female donkey.

DORSAL STRIPE: A dark line running from the top of the head to the top of the tail.

CROSS: A second line across a donkey's shoulders, intersecting the dorsal stripe to form a cross.

SPECTACLES: A term coined by the author to describe the pale rings often seen around a donkey's eyes.

UNDER-SADDLE: Riding.

HANDS HIGH (HH): Donkeys are measured at the withers the highest point where the neck meets the body. They are measured in "hands" (hh). One hand is 4 inches.

Foreword

Although donkeys—also known as asses—are members of the horse family, they are genetically different from horses themselves. During my research for this book, I learned the donkey genus includes many distinctive breeds.

I have also come to appreciate the historical roles played by these sturdy, adaptable creatures in almost every corner of the world. Whether they pulled a farmer's plow, carried the equipment for wars, were ridden or driven as transport or were simply kept for their milk, it is difficult to imagine where we might have been without them.

Sadly, the mechanization of farming in the twentieth century led to a reduced need for working donkeys. Today a number of traditional breeds are verging on extinction although breeders and governments are setting up programs to reverse this trend. Ever versatile, the donkey

has now found new vocations in the tourist industry and as valuable support in animal-assisted therapies.

Donkeys and mules are highly intelligent animals that have a great sense of self-preservation—which may seem like stubbornness. If they are unsure of what they are being asked to do, they just won't do it.

Jack Byard

Bradford, 2010

ABYSSINIAN

NATIVE TO
Ethiopia

SIZE
12hh on average (48")

A desert animal that survives on poor forage

The Abyssinian is tan to grey with a pale grey underbelly, inner thighs and muzzle. The ear-surround is dark, as are the tips of the mane. They have a dorsal stripe and cross.

The Abyssinian lives mainly in the desert areas that vary from below sea level to 2,300 feet above, and where rainfall seldom exceeds 8 inches a year. They have therefore developed the ability to survive on poor-quality grass and forage. They live in small groups of up to five animals with the number depending on the quantity of food and water available.

In some areas, Abyssinian donkeys are used for food which, added to the competition with domestic livestock for food and water, has caused a 90% reduction in numbers. Their milk is said to be good for skin conditions or for those who are lactose intolerant.

The breed is critically endangered. They are now used in protected areas for transporting all manner of goods, from food and water to fuel and building materials.

ABYSSINIAN

KNOW YOUR DONKEYS & MULES

AMERICAN MAMMOTH

NATIVE TO
USA

SIZE
There is no standard size

Big and strong, the most common donkey in America

The American Mammoth can be black, chestnut, spotted, dun, white, dapple-grey or palomino.

In America, before the days of mechanization, most of the power for transport and agriculture came from donkeys so the larger and stronger they were, the better. The native donkeys were in need of improvement and the European breeds chosen for this mammoth task (excuse the pun), were the Catalan, Andalusian, Majorcan and Maltese. By 1890, the demand for Majorcan (now Balear) donkeys was such that their native area was almost cleared of stock. George Washington is credited with developing the American Mammoth.

The coming of mechanization in transport and agriculture meant there was no longer a use for this sturdy, reliable giant. As a result many thousands were slaughtered. A few breeders were able to hold onto their stock so, although the breed is rare, it is safe from extinction.

AMERICAN MAMMOTH

KNOW YOUR DONKEYS & MULES

American Spotted

NATIVE TO
Probably Mexico

SIZE
9hh-16hh (36–64")

Can guard sheep and cattle

The American Spotted can be any donkey color as long as the main body color is broken with white patches that cover somewhere between 5% and 99% of its body.

It is widely believed the American Spotted donkey originated with the American Wild Burro. Prior to this, the American donkey was dark brown, black or roan. The Wild Burro was a smaller animal but with a much larger range of colors including, you guessed it, spotted. These two breeds were crossed to create the impressive American Spotted.

This lovely animal has been grazing the fields of America since the 1700s. In the early days, it was used entirely as a work animal and was considered unsuitable for any other purpose. The American Spotted is still used as a working animal today but is also used in harness, under saddle and to sire colorful mules.

Donkeys, including the American Spotted, have an aversion to intruders that can be put to good use for guarding sheep and cattle.

AMERICAN SPOTTED

KNOW YOUR DONKEYS & MULES

American Wild Burro

NATIVE TO
The deserts of North America

SIZE
11hh on average (44")

Favored by Gold Rush prospectors

American Wild Burros are normally grey but sometimes black or brown. Most of them have pale underbellies. They usually have a dorsal stripe and cross, leg stripes and a dark outline to the ears.

The Spanish introduced the Wild Burro into the southwest deserts of America in the sixteenth century. "Burro" is the traditional Mexican name for "donkey." As a pack animal they were in great demand in this arid countryside because of their ability to survive on low-quality forage and little water. The Burro can sustain a water loss of up to 30% of its body weight and replenish it with a 5-minute drink.

The gold and silver prospectors of old relied on this sturdy breed to carry supplies across the desert. In many cases the Burro would survive while the prospector would be beaten by the harsh desert conditions.

In 1971, Congress unanimously passed the Wild Free-Roaming Act, giving federal protection to the Wild Burro and Mustang populations and making it illegal to harass, brand, capture or kill these animals.

AMERICAN WILD BURRO

KNOW YOUR DONKEYS & MULES

AMIATA

NATIVE TO
Italy

SIZE
12.2hh–13hh (48–52")

**SURE-FOOTED
IN THE MOUNTAINS**

Sure-footed in the mountains

Amiata are mouse-grey with a pale grey underbelly and muzzle. They sometimes have white on the inner legs, throat and around the eyes. The ears have black edges and there is a well-defined dorsal stripe and cross. Dark stripes can occasionally be found on the legs.

Maritime traders introduced the Amiata donkey to the Amiata Mountains in Italy in about 2000 BC. Until the latter part of the nineteenth century they were a common sight in the Tuscan mountains. This wiry donkey was and still is, although in a much-reduced way, used on farms as a pack animal or for transport. It is an ideal working companion in these mountainous regions.

At the beginning of the twentieth century there were about 2,500 Amiata donkeys. Numbers declined for various reasons and reached an all-time low between 1970 and 1980. In 2003, only around 200 breeding Amiata survived. There is now an extensive breeding program in place and the situation is improving. The Amiata is used for work in inaccessible forest areas or to carry packs for walkers. Their milk is used in various beauty products.

AMIATA

KNOW YOUR DONKEYS & MULES

ANDALUSIAN

NATIVE TO
Spain

SIZE
13.2hh–15.1hh (52–60")

The oldest breed of European donkey

They are either grey-spotted, a shade of grey or strawberry roan. They rarely have the dorsal stripe and cross.

The Andalusian is thought by many to be the oldest European breed of donkey, having arrived in Spain more than 3,000 years ago. They are descendants of a now-extinct breed of large Egyptian donkey known as the Pharaoh.

The breed was mainly used on farms in the El Guadalquivir region of Spain. The farmers required a strong and docile animal that could work in the cork forests and citrus groves, and produce good mules.

The Andalusian has a good resistance to disease and adapts well to arid conditions. It has boundless energy and a long lifespan. Despite these excellent qualities, in the 1980s the breed almost became extinct, with the decline due largely to mechanization. The breed numbers were estimated at 120 to 150 worldwide. A number of organizations are now working to save this handsome, gentle animal from extinction.

ANDALUSIAN

KNOW YOUR DONKEYS & MULES

ASINARA

NATIVE TO
The Island of Asinara off the northwest coast of Sardinia.

SIZE
7.3hh–10.1hh

It's white but not an albino

The Asinara appears to be albino. It is white all over with pink skin visible through the hair. The muzzle has no hair and the hooves are without pigment. The irises are pink or blue. It is, however, only considered a "partial incomplete albino" because there are just a few grey animals.

It is believed the Asinara is descended from a group of white donkeys imported from Egypt in the 1800s by the Marquis de Mores (the Duke of Asinara). The breed is hardy and has the ability to live off meager grazing, which is very fortunate because fresh water and quality vegetation are in short supply on Asinara. This small, rugged animal manages to survive and multiply in this harsh environment.

Human inhabitants were forced to leave Asinara in the late-1800s and went to live in Stintino, on the coast of Sardinia, leaving the donkeys to fend for themselves. Asinara is now part of the Italian National Parks system and the animals are under the care and protection of the Sardinian Forestry Agency and the Faculty of Veterinary Medicine in Sassari.

ASINARA

BALEAR

NATIVE TO
Spain

SIZE
14.2hh–15hh (56–60")

Strong and healthy but somewhat nervous

Balears are black or almost black with a pale grey around the mouth, eyes, muzzle and underbelly.

It is believed by many that this European donkey has its origins in the western Mediterranean from where it spread into southern Europe. In 2005, the breed known as the Mallorquin or the Majorcan officially became known as the Balear. It is noted for being strong and healthy with a slightly nervous temperament. The females were used mainly as draft animals, carrying crops of olives, producing the power for the corn mills and for transporting the farmer and his family. The male was dedicated to mule breeding and used to produce the Kentucky mule.

As with most donkeys, numbers declined with the increase in mechanization and extinction was a distinct possibility. In 1990, a group of farmers and breeders led the fight back with a core of 50 animals and founded an organization, Asociacion de Criadores y Propietarios de Pura Raza Asnal Mallorquina (ACRIPROASMA) and created a stud book. The government recognized the group in 2002. The breed is now used in the conservation and maintenance of wood and scrubland.

21

BALEAR

KNOW YOUR DONKEYS & MULES

BOURBONNAIS

NATIVE TO
France

SIZE
13.1hh–14.1hh (52–56")

Needs a lot of water

The body of the Bourbonnais is beige to chocolate-brown while the muzzle and underbelly are grey, as are the spectacle markings around the eyes. The dorsal stripe and cross are a darker brown. The legs can have horizontal stripes.

The Bourbonnais comes from Allier in Auvergne, France, where it was depicted on a frieze in the church of St. Julien that dates from the early twelfth century. They helped in the maintenance of farms and vineyards, and were a big part of rural life in Allier, hauling supplies or produce and providing transport. In addition to its diet of grass, flowers and hay, the Bourbonnais drinks about 4 gallons of water a day.

This gentle donkey is healthy and long-lived: 30 to 40 years is normal, but reaching 50 is not exceptional. Though on the endangered list, the Bourbonnais is becoming a success in the tourist industry. This sure-footed animal is ideal for trekking over rough terrain but is just as much at home with gentler pursuits in harness or under saddle.

BOURBONNAIS

KNOW YOUR DONKEYS & MULES

CATALAN

NATIVE TO
Catalonia in northeast Spain

SIZE
13.3hh (53")

Hard working, long lived, notably amorous

The Catalan is black, turning to chestnut in the winter. They have a greyish-white underbelly, muzzle and spectacles.

The Catalan is a very old breed related to the Balear and the Zamorano Leones. It was once the main source of power on the farm. In its heyday, there were more than 50,000 at work, but at last count, numbers were down to 500, of which 400 were in Catalonia. A conservation program is now in place and the population is slowly increasing.

The Catalan is not known for its easy-going temperament but it is hard working, long-lived and remarkably disease resistant. To put it delicately, male Catalans are known for their enthusiasm for the female of the species. These positive attributes created a demand for its use in improving other breeds and it was used in the creation of the American Mammoth. The Catalan's ability as a draft and pack animal has won it many competitions and prizes.

CATALAN

KNOW YOUR DONKEYS & MULES

Cotentin

NATIVE TO
France

NOW FOUND
11.3hh–13.2hh (44"–52")

An ideal trekking companion

The Cotentin is ash-grey, blue-grey or dove-grey, occasionally with a reddish tint on the head. The underbelly, insides of the forelegs, thighs, muzzle and spectacle markings are a greyish-white. They have a dorsal stripe, cross and occasionally leg stripes.

For centuries the Cotentin (pronounced Ko-ton-tan), has grazed and worked on the small farms of Lower Normandy. Just over a century ago, the Manche area of Normandy had about 9,000 donkeys used mainly as pack animals for transporting milk, hay, manure and the apples for making cider.

Mechanization in agriculture meant the Cotentin has disappeared from the countryside today but all is not lost. The breed has found a niche in tourism, especially trekking holidays where this gentle and willing animal makes an ideal companion.

The Cotentin is also used for animal assisted therapy. This began in the UK in the 1700s and was championed by Florence Nightingale during the Crimean War because the sick and disabled can derive great benefit from the companionship of animals.

COTENTIN

KNOW YOUR DONKEYS & MULES

CYPRIOT

NATIVE TO
Cyprus

SIZE
13hh on average (52")

Still used for daily transport

Cypriots are black, brown or grey with a black mane. They have a lighter underbelly and a cream muzzle and spectacles. Some have the dorsal stripe and cross.

The Cypriot, or Karpas, donkey has a long history in Cyprus, a skeleton having been found in a tomb in the city of Enkomi. The ancient and the modern frequently meet on the roads of northern Cyprus where the donkey has not been entirely abandoned as transport. To own a donkey was once a luxury and apart from being the family transport, they were used to carry olives and cereals to the mills. The British army on the island also used them during World War II.

The population of the wild Cypriot donkey increased after the Turkish invasion in 1974, when the Greek Cypriots were forced to abandon their animals. The donkeys were eventually taken to the remote and beautiful area of Karpas in northeast Cyprus, where it was thought they would stand the best chance of survival.

CYPRIOT

KNOW YOUR DONKEYS & MULES

ENCARTACIONES

NATIVE TO
Spain

SIZE
11.3hh (47")

The muscle power for the Spanish explorer

The Encartaciones is black or chestnut with lighter areas around the mouth and eyes and a white underbelly. They have small hooves and ears and some animals have dark stripes on the legs. They have a dark dorsal stripe and a tuft at the end of the tail.

This docile donkey from the Basque region of Spain has been known in the area since the fifteenth century. It has been the muscle power for the Spanish explorer, the military and the small farmer who used it to shepherd stock between winter and summer grazing or as a draft animal. This small donkey, only 47 inches tall, is still used on small Basque farms although mechanization led to a dramatic fall in numbers and it is thought there are only 100 Encartaciones left today.

In 1996, the Association para la Defense del Burro Encartaciones (ADEBUEN) was created to protect the breed and prevent a further fall in numbers. These small, strong animals have for centuries been the power helping to create wealth for many. We cannot abandon them now.

ENCARTACIONES

KNOW YOUR DONKEYS & MULES

Grand Noir du Berry

NATIVE TO
France

SIZE
12.3hh–14.1hh (48"–56")

Ideal for tourists on holiday

The Grand Noir du Berry is black or a shade of brown. The underbelly, inside forelegs and thighs are a pale grey as are the spectacles.

In the mid-nineteenth century there were almost 9,000 Grand Noir du Berry and they could be seen working on small farms and vineyards where they were the draft animal of choice. It was about the same time that the Grand Noir replaced the human draft animal to haul barges along the Berry canal and onward to the Briare canal and Paris.

Within 100 years, mechanization took the Grand Noir to the brink of extinction. A small group of dedicated farmers in the Lignières-en-Berry, in conjunction with a campaign to save rural traditions, came to the rescue. By the year 2000, 1,000 Grand Noir du Berry were registered. Today the breed is mainly used in the leisure industry to carry equipment for tourists on trekking holidays. It is an ideal task for this strong yet docile animal.

GRAND NOIR DU BERRY

KNOW YOUR DONKEYS & MULES

IRISH

NATIVE TO
Ireland

SIZE
13hh on average (52")

Arrived with the Romans

Irish donkeys are mainly brown or black with a dorsal stripe and cross.

The Irish donkey is not recognized as a specific breed but this little donkey has played such an important role in Irish history and established itself so soundly in the country's affections that I thought it deserved a page of its own.

It is strongly believed that donkeys arrived in the British Isles with the Roman armies almost 2,000 years ago. During the Peninsula War in Iberia (1808-14) there was great demand for Irish horses. Donkeys from England were traded for these horses and records suggest this is when the donkey became more visible in Ireland.

One line of my research suggests that, in its early days in Ireland, the donkey was owned solely by the wealthy and used for milk. There is mention of a single donkey being taken as spoils at the fall of Maynooth Castle in 1534. In later years, the powerful Irish donkey was used for plowing, carrying, or as transport.

KNOW YOUR DONKEYS & MULES

KIANG

NATIVE TO
Tibet, China and India

SIZE
13.3hh (53")

Largest member of the wild ass family

Kiangs are a dark chestnut in winter and a sleek reddish-brown in summer. The underbelly is white, as are the legs except for a brown stripe down the front. The short brown mane stands up vertically. A dark stripe runs down the spine.

The Kiang is the largest member of the wild ass family. It has the appearance of a horse, but the long thin tail with a tuft at the end and hairs growing up at the side give it away. They graze the Tibetan plateau at heights from 8,200 to 17,400 feet above sea level, living mainly off the wispy, fine-textured stipa grass and other low-growing vegetation.

During August and September, they gain about 100 pounds in weight and a thick winter coat to help them survive the harsh winters where temperatures range from −15° to −31°F. Throughout the mating season, vicious battles take place between the reigning stallion and the wannabees. The Kiang has only two predators: man and wolf. When under attack the herd forms a circle, lowers their heads and kicks out violently.

KIANG

KNOW YOUR DONKEYS & MULES

MAJORERA

NATIVE TO
The Canary Islands

SIZE
10.3hh (42") on average

A born survivor in harsh conditions

The Majorera is grey with a pale grey muzzle, underbelly, inner ear and spectacles. The ears have dark edges. They have a dorsal stripe and cross.

The Majorera arrived in the Canary Islands from northwest Africa in the fifteenth century where it has adapted to the volcanic environment. Like most donkey breeds, it is a born survivor and endures harsh conditions with little food or water.

The Majorera is known for its vitality, health and strength. This gentle animal has been used, and still is to a lesser extent, as a pack animal, for pulling carts, plowing fields and under saddle. They are now mainly used in recreational pursuits and tourism. At the last count the population was fewer than 200 animals. Because it is in danger of extinction, it is now protected by the local agricultural administration. The Majorera can live at least 30 years, so owning one is not a short-term commitment.

MAJORERA

KNOW YOUR DONKEYS & MULES

Martina Franca

NATIVE TO
The Puglia region of Italy

SIZE
13.1hh–15.3hh (52"–62")

Strong and intelligent pack animal

The Martina Franca is dark brown. The underbelly, inner thighs and muzzle are grey.

According to local Puglia inhabitants, this large donkey is descended from the original local breed that was crossed with Catalonian donkeys imported into Puglia during the Spanish occupation.

Between the eighteenth and nineteenth centuries, the Martina Franca, then known as the Puglese, thrived. It was reared all over Italy and exported to France, Germany, Poland, Slovenia, Hungary, Greece, Brazil, Argentina, South Africa and India. It was not until 1904 that the Puglese was renamed the Martina Franca.

This strong and intelligent breed was used as a pack animal and to produce mules. Because of their ability to cope with poor pasture and rough terrain, they were used by alpine troops to carry equipment over the difficult mountain passes. After World War II increased mechanization meant that numbers began to fall, but there is now a breeding program to ensure the future of this historic animal.

MARTINA FRANCA

KNOW YOUR DONKEYS & MULES

MARY

aka Maryiskaya

NATIVE TO
Turkmenistan, central Asia and surrounding areas

SIZE
11.2hh–15.3hh (45"–62")

Steady progress on steep paths

The Mary is most frequently grey but occasionally black. The underbelly is grey or white as are the insides of the forelegs and thighs. The legs have horizontal dark stripes. Most Marys have a dorsal stripe and cross.

Information about the origins of the Mary (as it is known in western countries) donkey is far from clear. Today, the Mary is bred largely in the Mary province and Ashkhabad region of Turkmenistan. It is extremely adaptable and able to survive in harsh mountainous regions or the lush regions of the plains. Their native countryside is harsh and inaccessible so it is necessary to transport goods over steep slopes and mountain tracks. The Mary walks with a short pace, which reduces the swaying of a load and allows for steady progress on steep paths.

An improving rural economy and increased mechanization led to a serious drop in numbers. However, the Mary is still used for expeditions and mountain rescue parties and a breeding program has been put in place on reserves in the area to ensure the survival of this ancient breed.

KNOW YOUR DONKEYS & MULES

MEDITERRANEAN MINIATURE

NATIVE TO
Sicily and Sardinia

SIZE
8.1hh–8.2hh (32")

Friendly and intelligent, a great pet

The Mediterranean Miniature is mainly grey-dun or a shade of grey. Less common colors are spotted, chestnut, white, dark brown and black. They have a lighter colored underbelly and muzzle. The ears have a darker edging and there is a dark tip to the tail. The majority has the dorsal stripe and cross.

The Mediterranean Miniature is a naturally small donkey rather, not bred down from a standard breed. They were once used as a draft animal and for grinding grain in peasants' houses. Blindfolded and harnessed to the mill, they would walk for hours in endless circles. History has not always treated donkeys kindly.

Today this friendly and intelligent donkey is kept as a pet. It makes a good companion animal for children or people with disabilities and is a welcome visitor in many nursing homes. They enjoy human company and will nudge you to gain attention. Many Mediterraneans are used to carry packs on camping trips and they are easily trained to pull a small cart. The Mediterranean can live for 40 years.

MEDITERRANEAN MINIATURE

KNOW YOUR DONKEYS & MULES

NORMANDY

NATIVE TO
France

SIZE
9hh–12.1hh (36"–48")

Small but so strong it can carry its own weight

The Normandy is brown and occasionally mouse-grey. The underbelly, inside forelegs, thighs and muzzle are a greyish-white. The spectacle markings around the eyes sometimes have a hint of red. They have a dorsal stripe and cross.

The Normandy, like most French donkeys, worked in agriculture as a pack animal, carrying corn or hay and taking metal milk churns out to the fields at milking time. In 1997, there were 4,500 registered with the Ministry of Agriculture. It was also used in market gardening to carry produce to local markets. This small yet strong animal is capable of carrying the equivalent of its own weight, usually around 400 pounds.

Today, the Normandy is frequently used under saddle or in harness in their new role at seaside resorts, village festivals and in the tourist industry—not forgetting their popularity as pets.

The Association of Normandy Donkeys ensures the natural strength and calm temperament of the breed is not undermined by careless breeding.

NORMANDY

KNOW YOUR DONKEYS & MULES

ONAGER

NATIVE TO
Mongolia, northern Iran, Tibet, India and Pakistan

SIZE
12hh on average (48")

Very fast and too wild to domesticate

The Onager is reddish-brown in the summer and yellowish-brown in the winter. They have a black stripe with a white border down the middle of their back. The underbelly is buff, they have a dark upright mane, the ears have dark tips and the tail ends in a dark tassel. The legs are short.

The Onager is a descendant of the Asiatic Wild Ass. It is small and slender compared to many donkeys but what it lacks in size it makes up for in speed and endurance, being capable of a constant 40 mph over a 15-mile stretch. The world's expert on the breed told me that despite tales to the contrary, the Onager has never been domesticated. In Hebrew it is called Perah, meaning "wild."

They live in small groups of five or six in semi-desert that can reach temperatures of 120°F. They feed on grasses, herbs and shrubs and never move too far from a watering hole. The Onager is now an endangered species and in 2005, numbers were estimated at around 600.

ONAGER

KNOW YOUR DONKEYS & MULES

PANTELLERIA

NATIVE TO
Sicily

SIZE
11.3hh-13.3hh (45"–54")

Slender, elegant and graceful

Pantellerias are black or dark bay. The short, glossy coat has an oily feel that is unusual in a donkey. The muzzle is a pale grey, as are the underbelly and the inner parts of the legs. The tail is bushy.

The Pantelleria donkey dates back to the first century BC and is the result of crossing African and Sicilian breeds. I have heard them described as "slender, elegant and graceful with a gait reminiscent of a camel." The Pantelleria was important in the growth and development of Sicily and they were so valued by the peasant farmers that each donkey was given its own stable.

About 20 years ago, the last Pantelleria drowned. A group of nine donkeys were selected and bred in order to remove unwanted genes and recreate the breed. Soon they will be returned to their natural habitat, where they will work in a protected area as a draft animal, as tourist transport or to fulfill the ever-growing demand for donkeys in animal assisted therapy.

KNOW YOUR DONKEYS & MULES

POITOU

NATIVE TO
France

SIZE
13.1hh–14.3hh

Almost extinct but the future is bright

The Poitou is dark brown or black with a naturally long, shaggy coat. They have a white underbelly, nose and spectacle markings with no stripe or cross.

The Poitou (pronounced pwa-too) is officially called Le Baudet de Poitou and considered one of the rarest donkey breeds. Historians believe this gentle donkey was grazing in France 2,500 years ago. In the Middle Ages, to own a Poitou put you among the nobility. The main use of the Poitou was for breeding. They were crossed with the Trait du Poitevin, a French cob, to produce the Poitou mule used in agriculture. At its peak, the region was known to produce in excess of 18,000 Poitou mules per year.

Mechanization meant the Poitou was no longer needed and by 1977, only 44 remained worldwide. Extinction was a possibility. At this point, a collection of authorities and enthusiasts came to the rescue of this easy-going donkey. The latest figures show there are now 400 worldwide, of which 180 are pure bred. The future is definitely looking brighter for the Poitou.

POITOU

KNOW YOUR DONKEYS & MULES

PROVENCE

NATIVE TO
France

SIZE
12.3hh–13.1hh (48"–52")

Docile and sure-footed, ideal in mountains

The Provence has a dove-grey coat that sometimes has a hint of red or brown. The underbelly, inner forelegs, thighs and muzzle are white. Occasionally they also have a white jaw line and spectacle markings. The forehead, ears and the edge of the eyes usually have a brown-red tint. The legs can have black horizontal stripes. The dorsal stripe and cross are well defined.

The Provence donkey dates back to the fifteenth century and was named after the region where it was created. The shepherds of the area developed and improved the Provence until it was ideal for their needs. Their legs are strong and their hooves are slightly larger than those of other breeds, making this docile, sure-footed animal ideal for carrying heavy loads on treacherous mountain paths.

In the late-nineteenth century, there were a recorded 13,000 Provence donkeys, but by 1993, only 330 existed. The population has now reached 600, which is an improvement—but there is still a long way to go before the Provence is safe from extinction.

PROVENCE

KNOW YOUR DONKEYS & MULES

PYRENEES

NATIVE TO
The Pyrenees in France and Spain

SIZE
11.3hh–13.1hh (45" to 52")

Can carry gear for four people

The Pyrenees is mainly black but can also be brown or chestnut. The underbelly is pale grey as are the insides of the forelegs and thighs. This donkey does not have a cross on its back.

The Pyrenees donkey is strong, graceful and intelligent. For a small farm with poor land, a donkey was the ideal animal to cope with harsh conditions and poor grazing. The Pyrenees was an important partner and worked as a draft animal or in harness, delivering bread or cheese, transporting hay and firewood and bringing ice down from the snowfields.

Until the early years of the twentieth century, the Pyrenees donkey flourished, but much as with other donkey breeds, numbers declined and in 1990, the numbers reached an all- time low. Today the leisure and tourist industry is creating an ideal opportunity for this hard-working animal, transporting tourists to beautiful but otherwise inaccessible areas. It has the ability to carry a quarter of its own body weight so will carry the camping equipment of four people. That is my kind of camping companion.

PYRENEES

KNOW YOUR DONKEYS & MULES

RAGUSANO

NATIVE TO
Sicily

SIZE
13.1hh–14.1hh (52"–56")

Milk is suitable for those with food allergies

The Ragusano is dark bay with a pale grey muzzle and spectacles. The underbelly is a lighter brown and the mane and tail are black.

The Ragusano is a newly established breed created by crossing the Pantelleria, the Martina Franca and the Catalan. It was officially recognized in 1953 by the l'Instituto di Incremento Ippico di Catania, which keeps the population records and has laid down the breed standard.

In the early days, the Ragusano's ancestors would be used in agriculture and for breeding mules, which were ideal for winding mountain paths and thus were frequently used by the military. The Ragusano has all the attributes of its ancestors and can live for up to 45 years.

In recent years, donkey's milk has been rediscovered as an anallergic (not allergic) food that is almost identical to human milk and has been found suitable for children with food allergies. The milk is also used to produce creams, soaps and moisturizers for the beauty industry.

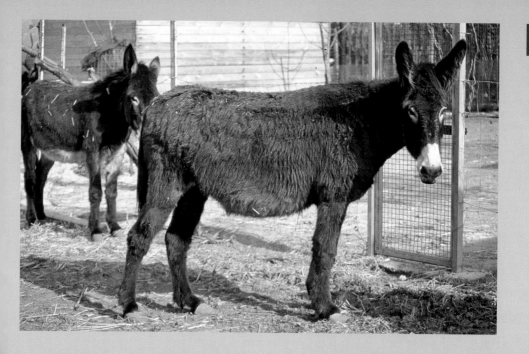

KNOW YOUR DONKEYS & MULES

ROMAGNOLO

NATIVE TO
Emilia-Romagna in the Forli province of Italy

SIZE
13.2hh-14.1hh (52"–56")

Strong, willing, reliable, and lively

Romagnoloes are dark bay with a pale grey underbelly and muzzle. There is white on the inner legs, throat and inside the ears. They have a dorsal stripe and cross and dark stripes are occasionally found on the legs.

The Romagnolo donkey is the native breed of Emilia Romagno in Italy and dates back to the fifth century. It was once widely used as a pack animal, under saddle and in agriculture because the animals are strong, willing and reliable with a lively character.

One of the major causes of the decline in numbers was the depopulation of the area. The existing workforce in agriculture aged while the younger generations left to find work. Demand for the Romagonlo decreased. The decline continued into the 1970s when the Romagnolo was threatened with extinction. There is now a breeding program in place and the Romangolo works maintaining forests in the mountainous marginal areas, which are difficult to use for agriculture but are important to keep the ecosystem in balance.

KNOW YOUR DONKEYS & MULES

SARDINIAN

NATIVE TO
Sardinia

SIZE
7.3hh–10.3hh (28"–40")

Carried water and farm equipment in the mountains

The most common color of Sardinian is grey with a lighter underbelly, inner thighs and muzzle. The edges of the ears are dark, as are the tips of the mane. They have a dorsal stripe, cross and horizontal dark stripes on their legs.

Much of the early history of the gentle Sardinian donkey is known through the writings of the Jesuit Priest Francesco Cetti in 1774 and the breed has changed very little since then. Do not be fooled by its diminutive size. This is a strong and hard-working animal that is ideal for use in steep, narrow passages and hills, yet strong enough to carry agricultural equipment and water to towns and villages.

The Sardinian donkey has been working on the island for more than 2,500 years but for the last 20 years has been considered critically endangered by worldwide animal welfare lists. Numbers of the purebred Sardinian are still falling. In 1965, there were 27,000 registered, but in 1995, there were only 150. The increase of cars and general mechanization are the main reasons for the decline.

SARDINIAN

KNOW YOUR DONKEYS & MULES

SOMALI WILD ASS

NATIVE TO
The southern Red Sea area of Eritrea and the Afar region of Ethiopia and Somalia

SIZE
12hh–14.1hh (48"–56")

Wild ancestor of the modern donkey

The Somali Wild Ass has a short, smooth coat that is light grey to fawn and shading to white under the body and on the legs. The mane is thick and upright with black tips, and the ears are large with fluffy black edges. The legs have horizontal black or dark brown stripes like a zebra. The tail has a tuft at the end.

The Somali Wild Ass is a sub-specie of the African Wild Ass, which in turn is thought to be the wild ancestor of our domestic donkeys. They graze the wild, low scrubland and rocky terrain of eastern North Africa wherever there is access to surface water. The Somali Wild Ass tends to be a solitary animal primarily due to lack of food. The small herds that do gather tend to be females and young.

Today they are a critically endangered species with fewer than 300 animals in their native land. In years past, the killing of a Somali Wild Ass was a serious crime and it was decreed the killer's hand would be cut off.

SOMALI WILD ASS

KNOW YOUR DONKEYS & MULES

ZAMORANO LEONES

NATIVE TO
Northern Spain

SIZE
13.1hh–15hh (52"–60")

Its long winter coat makes warm blankets

Zamorano Leones range from brown to black. The underbelly, inner thighs, spectacles and muzzle are pale grey. The winter coat is coarse, thick and long while the summer coat is short and smooth. The ears can have long light-colored hairs hanging from the front.

This powerful donkey is derived from crossing the original local Spanish breeds with those imported from the Catalan region during the sixteenth century. The breed was originally used for farming and mule breeding and its long winter coat was used to make blankets. It adapts well to marginal land where grazing tends to be poor.

The Zamorano Leones suffered from mechanization in farming and numbers decreased dramatically. In 1982, it was classified as endangered. The Spanish military has a number of these donkeys and it is hoped that with further research and help from the European Union, disaster may be avoided.

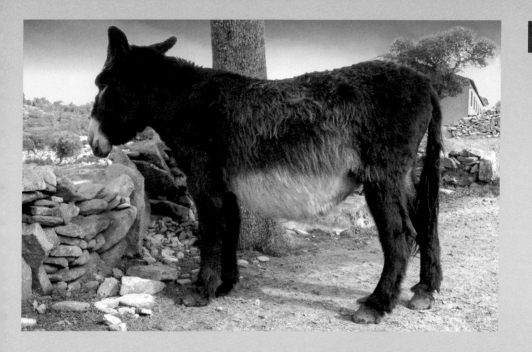

ZAMORANO LEONES

KNOW YOUR DONKEYS & MULES

ZEDONK/ ZONKEY

NATIVE TO
Most major continents

SIZE
Varies

Cross between a donkey and a zebra

The Zedonk or Zonkey is mainly grey or brown with zebra stripes.

The Zedonk is a hybrid, the cross between a donkey and a zebra. I am reliably informed that a Zedonk is the cross between a female zebra and male donkey while the offspring of a male zebra and a female donkey is known as a Zonkey. These hybrids are usually sterile and seldom occur in the wild.

The Zedonk pictured is a past resident of Colchester Zoo in England and the result of a female black ass and a passing zebra.

Other zebra hybrids include a zebrula/zorse—the cross between a female horse and a male zebra, and a zibrinny—the cross between a male horse and a female zebra.

ZEDONK/ZONKEY

KNOW YOUR DONKEYS & MULES

WHAT IS A MULE?
Dad is a donkey, Mom's a horse

According to the popular saying, "Mules can do anything a horse can do and they usually do it better and with a sense of humor."

The mule is a cross between a male donkey and a female horse. Mules come in all sizes and can be almost any horse color except pinto. The offspring of the complementary breeding, between a female donkey and a male horse, is called a hinny. All male mules and most female mules are infertile.

Mules are not, as is commonly believed, stubborn or stupid but highly intelligent, rugged and sure-footed. High on a mule's list of priorities is self-preservation, which I suppose could be looked upon as stubbornness. A mule is able to work in extreme climates that would be well beyond the capabilities of a horse.

George Washington is said to be responsible for having bred the first mules in America using an Andalusian jack, which was a gift from the king of Spain. This jack helped create the 58 mules George Washington had at work on his farm. They went on to become the most popular working animal in America.

Saddle Mule

NATIVE TO
USA

SIZE
14.3hh (58")

COLOR
Black with four white socks

Suitable for riding or driving

Pictured opposite is Eureka. Her father was a black Catalan cross American Mammoth donkey and her mother was a black-and-white paint horse. Mules of her type would be suitable for riding or driving.

The size and strength of a mule depends on the mare. The self-preservation instinct, inherent in all mules and the sure-footed gait are inherited from the jack. In Spain and Italy, pairs of Catalan-bred mules are frequently used for carriage work and will usually cost more than a pair of horses.

Eureka is the great-great-grand-daughter of Secretariat, the famous Kentucky Derby winner.

SADDLE MULE

KNOW YOUR DONKEYS & MULES

BELGIAN DRAFT MULE

NATIVE TO
USA

SIZE
To become 15hh (60")

COLOR
Chestnut with four white socks and a white rump

Sure footed and easy going in extreme conditions

Pictured opposite is Suburban. His father was a red-roan American Mammoth donkey and his mother was a Belgian Draft horse.

The Belgian Draft Mule is known for its sure-footedness and easy-going temperament. They can stand heat better than a horse, which is why they are used in Death Valley in southwestern USA. They are also the only animals used to haul people and supplies to the bottom of the Grand Canyon because they can cope with the very steep and narrow trails along sheer cliffs. They are extremely intelligent and affectionate.

Mule Racing
Mules are sprinters not distance runners

Today there about 70 racing mules in America even though the sport is little known outside California. A mule is a better sprinter than a long-distance runner and over a short distance, they can outpace Arabian and Appaloosa horses. Mule races normally take place at the end of horse racing events after the thoroughbreds and quarter horses have run. One of the most popular mule races is Winnemucca Mule Race in Nevada.

The best racing mules are female and the best of the best is Idaho Gem, born on May 4, 2003. Unlike horses, mules cannot be put out to stud; they are sterile and so unable to pass on their genes to a future generation. Instead, Idaho Gem is the very first cloned equine.

During the research into cloning Idaho Gem, Professor Gordon Woods of the University of Idaho discovered his work had great significance in learning the causes of certain male cancers. A company has been set up to carry out further research.

Acknowledgments

I would like to thank the following organizations without whose help *Know Your Donkeys & Mules* would never have been written.

The Donkey Sanctuary and NEDDI in England, The Donkey Sanctuary in Cyprus, El Refugio Del Burrito in Spain and Il Rifugio Degli Asinelli in Italy.

Many thanks also go to Tony Harman of Maple Leaf Images in Skipton for his advice on the finer points of photographs and the many people throughout the USA, Canada, Britain and Europe for all their assistance with pictures and information.

Thanks finally to Rebecca who keeps Granddad's feet firmly on the ground.

Any mistakes are mine and mine alone.

Jack Byard
Bradford, 2010

Picture Credits

(1) Don and Michelle Owens, Four Spokes Ranch, (2) The American Donkey and Mule Society, (3) Frank and Betsy Burke of www.lazybspotteddonkeys.com in Texas, (4) Nancy Kerson, mustangs4us.com, (5) Mr. G. Catillo, (6) El Refugio del Burrito, (7) Daniele Bigi, (8) Associació de Criadors de Pura Raça Asenca de les Illes Balears, (9) Ursula Bivans, (10) David Gaya, (11) Grosbois F. Haras Nationaux, (12) Izzet Zorlu, (13) Association para la Defense del Burro, (14) Rivalain Y. Haras Natiobaux, (15) Helen McCann, (16) Jan Morse, (17) SOO… Grupo para la Conservación Formento del Burro Majorero, (18) Terra Degli Asini–Lissaro, Italy, (19) Food and Agricultural Organization of the United Nations, (20) Penny Cooke, (21) Grosbois F. Haras Nationaux, (22) Netanel Nickalls, (23) Archivio RARE, (24) Annie Pollock, (25) The Donkey Sanctuary, (26) Ursula Bivans, (27) Archivio RARE, (28) Fattorie Faggioli, (29) Daniele Bigi, (30) Marwell Wildlife, (31) El Refugio del Burrito, (32) Bainbridge's Long Ears Acres, (33) Knute's Kustom Mule Kompany, Rod and Becky Knutson (34) Knute's Kustom Mule Kompany, Rod and Becky Knutson.

Discover these other great books from Fox Chapel Publishing

Art of the Chicken Coop
A Fun and Essential Guide to Housing Your Peeps
By Chris Gleason

A fresh approach to designing and building chicken coops with seven stylish designs that your flock will adore and your neighbors will envy.

ISBN: 978-1-56523-542-7
$19.95 · 160 Pages

The Chicken Handbook
A Practical Guide to Keeping Hens and Other Fine-Feathered Friends
By Vivian Head

All the fundamental information a backyard farmer needs to raise not only chickens, but geese, ducks, and turkeys too.

ISBN: 978-1-56523-686-8
$11.95 • 160 Pages

The Beekeeping Handbook
A Practical Apiary Guide for the Yard, Garden, and Rooftop
By Vivian Head

Provides the backyard farmer and honey artisan with all the fundamental knowledge needed to take the sting out of beekeeping.

ISBN: 978-1-56523-681-3
$11.95 • 160 Pages

Look For These Books at Your Local Bookstore or Specialty Retailer

To order direct, call **800-457-9112** or visit *www.FoxChapelPublishing.com*

By mail, please send check or money order to:
Fox Chapel Publishing, 1970 Broad Street, East Petersburg, PA 17520

More KNOW YOUR ANIMALS Books

Know Your Pigs
By Jack Byard

Twenty-eight breeds of pigs—from the American Guinea Hog to the Wild Boar.

ISBN: 978-1-56523-611-0
$6.95 • 64 Pages

Know Your Cows
By Jack Byard

Forty-four breeds of cattle—many you will recognize, but some are very rare.

ISBN: 978-1-56523-613-4
$6.95 • 96 Pages

Know Your Chickens
By Jack Byard

Forty-four breeds of chicken—some rare, each with a rich diversity of color, size and feather pattern.

ISBN: 978-1-56523-612-7
$6.95 • 96 Pages

Look For These Books at Your Local Bookstore or Specialty Retailer

To order direct, call **800-457-9112** or visit *www.FoxChapelPublishing.com*

By mail, please send check or money order to:
Fox Chapel Publishing, 1970 Broad Street, East Petersburg, PA 17520

# Item	Shipping Rate	
1 Item	$3.99 US	$8.98 CAN
Each Additional	.99 US	$3.99 CAN

International Orders - please email info@foxchapelpublishing.com or visit our website for actual shipping costs.